NINJA ST

Title: Ninja Style

ISBN-13: 978-1-942825-15-9

Author: Kambiz Mostofizadeh

Publisher: Mikazuki Publishing House

Copyright: 2017. All Rights Reserved.

Description: Ninja Style is a non-fiction book title about the Ninja of ancient Japan and the skills they used to defeat their opponents.

NINJA STYLE

TABLE OF CONTENTS

Where They Started – 3

Skills Vital to Ninja – 26

Offensive Weapons – 33

Koga Ryu - 47

Illusions – 74

Meditation – 81

Ninjas as Strategists – 103

Ninja Religion – 113

Ninja Culture – 124

Do Ninjas Exist – 132

Is Ninjitsu Obsolete – 144

Conclusion - 152

Catalogue - 167

NINJA STYLE

忍者

WHERE THEY STARTED

Some modern Ninja in the U.S. attempt to claim some type of mastery over the knowledge of Ninjitsu but it is safe to say that their understanding has only been surface deep as the Japanese have been wholly reluctant to share the actual secrets of their ancient art with the West. The last time a real Ninja existed was 400 years ago. Since 1603, the Ninja have ceased to exist in actuality. What is known about Ninja now has been gathered by hearsay or contact with descendants

NINJA STYLE

of Ninjas. Many believed that Ninjitsu in Japan began with Daisuke Togakure, a Ronin or masterless Samurai that had renounced his Samurai title after being part of a losing battle. During his time wandering as a Ronin, he met a Chinese Warrior-Monk named Kain Doshi in 1162. Togakure learned the secrets of Ninjitsu from Kain Doshi. The Ninja were a class of samurai that specialized in the deadly and silent art of Ninjitsu. But it is not really known who was or was not a Ninja. In many instances, the grandchildren would discover their ancestor was a Ninja 100 years after that person was born. Ninja were only efficient at their trade because of their ability to conceal the fact that

NINJA STYLE

NINJA STYLE

they were a Ninja. Most contracts between Daimyo and Ninja were oral and the written contracts that did exist did not reveal what the person would be doing for the Daimyo or feudal warlord. Ninjitsu is not a Martial Art or a way though it incorporates spiritual practices. Ninjitsu is a set of applicable techniques for infiltration, surveillance, and attack. Up until the late 14th century, ninjas were viewed by authorities as outlaws to be arrested and sentenced. Ninja clans were referred to as murder gangs. It was from the late 14th century during the Muromachi shogunate, that Ninjas served as titled landowners and took a more direct role. The 15th and 16th centuries were the peak of Ninja activity. The Sengoku Jedai

NINJA STYLE

Period or the Warring States period in Japan from 1467 to 1573, made the Ninja infamous throughout Japan and the entire world. It is known as the Warring States period but it should really be known as the Warring Clans period. It was multiple Samurai Clans that were fighting each other and the various Daimyo or Warlords in charge of those Clans made treaties and broke them to gain advantage. Refugees flooded the Japanese countryside, fleeing from war torn areas affected by the Samurai classes vying for power. The Samurai lived by a strict Code of Bushido that was honor based and was chivalric. In comparison, the Ninja used to be Samurai that had been made Ronin or had become rebels. The Ninja did not live by the

NINJA STYLE

same code of Bushido that the
Samurai did. The Ninja carried out
assassinations, surveillance, and
wore disguises to infiltrate an
enemy's camp. These things would
have been viewed as being
dishonorable by Samurai so they
contracted out such work to Ninja
Clans. The Ninja were skilled in
silent movement, attack, influencing
public opinion, and strategy. The
Ninja took silent movement so
seriously that they had special types
of shoes for infiltration that were
wooden sandals with animal fur
attached to deafen the sound when
walking on wooden floors. During the
Edo period in Japan in 1603,
Tokugawa Ieyasu brought all of the
Ninja to serve him, allowing him to
establish a vast network across

NINJA STYLE

Japan. But by doing so, he was able to vastly diminish the strength of the Ninja clans that were up until then operating as separate mercenaries for hire. Before 1603, the Ninja were simply referred to as Shinobi. Ninjitsu (or Ninjutsu) was referred to as Shinobijutsu. Shinobi or "stealer-ins" were named so because of their stealth techniques that allowed them to penetrate deep in to an enemy's castle without being detected, pulling off seemingly magical actions that had been unheard of by Samurai. Rumors began to circulate of the Ninja's seemingly invincible powers of invisibility. The Ninja evolved out of the Samurai classes but their specialty in stealth attacks, swift methods of disposing of opponents in hand to hand combat, and ability

NINJA STYLE

to seemingly disappear at will gave them the reputation of having almost magical powers. The Ninja were first and foremost masters of techniques of illusion or Genjutsu. Their ability to disappear was based on their advanced understanding of misdirection and techniques related to illusion. Although the Ninja acted in a mercenary like manner, they operated under the warrior code of Bushido. Ninjitsu was used to stop an invasion by an enemy, prevent the loss of life, or in the service of a Daimyo (feudal landowner). Ninjitsu that was used for personal reasons

NINJA STYLE

Chinese General and Master Strategist Sun Tzu was studied by the Ninja for greater insight in to strategy and tactics.

NINJA STYLE

was seen as morally weak and doomed to failure. Ninja clans were organized with the purpose of contracting their service to various Daimyo. Individual Ninja that were hired had to sign a secret contract with their Daimyo that would not reveal what their job activities were. Individual Ninja were sworn to secrecy to never reveal what they did for the Daimyo.

"Hostile armies may face each other for years, striving for the victory which is decided in a single day. This being so, to remain in ignorance of the enemy's condition simply because one grudges the outlay of a hundred ounces of silver in honors and emoluments, is the height of inhumanity. One who acts

NINJA STYLE

thus is no leader of men, no present help to his sovereign, no master of victory. Thus, what enables the wise sovereign and the good general to strike and conquer, and achieve things beyond the reach of ordinary men, is foreknowledge. Now this foreknowledge cannot be excited from spirits; it cannot be obtained inductively from experience, nor by any deductive calculation. Knowledge of the enemy's dispositions can only be obtained from other men."

-   Sun Tzu

Ninja Clans spent most of their time in training, when they were not conducting missions. Ninja missions, were designed to be stealthy and rapid. Most of the time spent by a

NINJA STYLE

Ninja was conducting reconnaissance and gathering information that would allow for rapid insertion and escape. The Ninja were not a blunt instrument for warfare as many popular Hollywood movies depicts, but rather a surgical strike force that is similar to modern military Special Forces. Ninjas were trained in the use of various disguises and using these disguises they were able to penetrate and live among their target for months at a time. The greatest and easiest disguise available to Ninja was that of a peasant or villager. Whenever Daimyo were under threat of being attacked, they would send officers to recruit villagers in to the Daimyo's army. The easiest and fastest place for the Daimyo's officers to recruit

NINJA STYLE

from was the village closest to the Daimyo's castle. Ninja would use simple and straight forward disguises that would allow them to blend in to a village for months at a time before being recruited in to an enemy Daimyo's army. This method would allow Ninja to walk in to an enemy's castle in broad daylight, unopposed and unnoticed. In many cases, Ninjas would scale a 10-meter-tall castle

NINJA STYLE

wall and walk right in to the castle without being seen, after dispatching a guard on the castle wall. The months and sometimes years of training a Ninja would undergo made them the deadliest warrior that the world had witnessed. Ninjitsu is widely believed to have started from Iga Province in Japan. The Ninja of Iga and Koga were famous and known to be greater, skill wise to all other Ninja. Iga and Koga were close to Kyoto, the center of power of Japan. The Iga Ninja were believed to have originally been Yamabushi or Warrior Monks. The Ikko-Ikki Uprising in the 15th and 16th century by Warrior Monks, Farmers, lower ranked nobles, and peasants against the Samurai class led to the invasion of Iga by Oda Nobunaga. They were

NINJA STYLE

able to maintain their independence free of the central government as their political and military power was feared as being able to rival even the central government. According to the Ninja Museum of Iga-Ryu "Iga was mostly manor lands, the people of Iga created living areas by manor in units of clans, formed an organized party of landowning farmers, and did not defer to the control of central regimes, an important 12-member council (representatives) was chosen from among the 50-60 members of the party in Iga, and they maintained safety in Iga by cooperation." The Iga-Ryu which pioneered Ninjitsu, was famous for its techniques that influenced all other schools in Japan. The three main book of the Ninja are the Shinobi Hiden, Mansen Shukai,

NINJA STYLE

and the Shoninki. The Shinobi Hiden
was authored by Hattori Hanzo, a
16th century Iga Ninja leader that
saved Tokugawa Ieyasu after Oda
Nobunaga's death. In one famous
battle between Tokugawa and
Takeda, Tokugawa was forced to
retreat to his castle after taking
heavy losses on the battlefield.
Tokugawa had only five men left
after this battle so he charged out
beating a drum which saved him
some time because the Takeda
thought that this was a trap. In the
middle of the night, Hattori Hanzo
and his Ninjas crept in to the Takeda
camp and wreaked havoc on them,
causing the Takeda army encamped
outside Tokugawa castle to flee.
Hattori has been attributed with
magical powers and referred to as

NINJA STYLE

being a Demon, but it was his perfect execution of operations and his meticulous planning rather than his fighting prowess, that gave him this nickname. Before his invasion of Iga Republic, Oda had steadily built up his forces in preparation and had continue his conquest of nearby provinces as he sought to use military force to unite Japan. Oda Nobunaga sought to destroy the Buddhist religious centers that served as the bases of power for the people of Iga. At that time, the monasteries had as much power as the central government. Since Iga Republic had resisted assimilation and had sought independence by uniting its farm owners and monasteries, the content people of Iga gave favor to the Iga Ikko-Ikki

NINJA STYLE

Republic and resisted the Central gov't. In 1581, Daimyo Oda Nobunaga invaded and encircled the Iga Ikko-Ikki Republic with 50,000 soldiers. Oda's army burned entire villages down to the ground. In addition, Oda had nearly 600 Ninjas that were working for him during the invasion. The war lasted 3 years and resulted in Iga being brought under the power of the Japanese Central Government. As the Ninja of Iga were masters of guerrilla warfare, they were able to wage a campaign of attrition that gave many losses and worries to Oda's commanders, but the Ninja forces of Iga were no match in conventional warfare with Oda's highly disciplined and vastly larger army. Many poor villagers, men, women, and children were

NINJA STYLE

slaughtered in Iga province, simply because of where they lived. Such was the brutality during the Sengoku Jedai period. The Samurai class had a problem of constantly shifting loyalty during this time and it was not uncommon for a Samurai commander with thousands of troops under his command to change his allegiance and loyalty over monetary gain. The Samurai, although immortalized in film and literature as being the ultimate symbol of loyalty, were in fact changing their loyalty so constantly that Daimyo had to keep their plans compartmentalized to prevent being betrayed. The Ninja were far more loyal than Samurai were and that is because the Ninja worked for contracts that included financial compensation. The Samurai

NINJA STYLE

were quick to go to war over a perceived sleight or a perceived sign of disrespect, but the Ninja held no such false motives. The Ninja based their loyalty on the contracts they received. In many instances, Ninja could have betrayed their paymasters and turned over Daimyo to be arrested and killed. But the Ninja, were able to maintain their reputation of being loyal by precisely avoiding the Machiavellian style of politics which the Samurai actively engaged. Intrigue and subterfuge played a huge role in the life of Samurai despite their overt stance against it. The Ninja were far more skilled in this activity but they did it for contracts and money, making them far more loyal and dedicated than the Samurai. The Samurai may

NINJA STYLE

have practiced the code of Bushido and loyalty among lower levels of Samurai and among Samurai retainers (valets), but the Daimyo hardly practiced loyalty and changed their loyalty to whomever they believed would be able to raise their status and power in Japan. Contracts between Ninja Clans like the Iga Ryu and Koga Ryu were always honored and loyalty was unquestionable. The Daimyo turned to Ninja in many cases because they could not trust their own Samurai to carry out vital tasks related to their Clan's security. Samurai were a blunt instrument of warfare that were called out in dire situations. Samurai and the various foot soldiers that made up a Clan's army were greatly expensive to arm, expensive to

NINJA STYLE

maintain, and even more expensive to mobilize. The Ninja could be mobilized on a short notice via contract and payment without the visual noise created by the amassing of a large military force. The Ninja could be mobilized to assassinate enemies without giving off the signal that an attack was impending. The silence of the Ninja is what created the greatest alarm for those that could be potentially targeted for assassination. An attack by Ninja from a particular Daimyo would carry no forbearing, no warning, and no signs that such an attack could materialize. In contrast to the Samurai who were a blunt instrument of warfare, the silent approach of the Ninja created a permanent and ever-present fear

NINJA STYLE

among Daimyo and they all, in one
form or another, maintained some
type of Ninja patrol that actively
walked through their forts and
castles searching for hidden
assassins.

NINJA STYLE

SKILLS VITAL TO NINJAS

The Ninjas spent much of their time training and honing their skills and these included:

1. Hand to Hand Combat
2. Silent Movement
3. Breaking Locks
4. Infiltration Techniques
5. Escape Techniques
6. Invisibility Techniques
7. Con-Artistry & Lying
8. Disguise Creation
9. Illusions/Misdirection
10. Explosives/Smoke Grenades
11. Meditation
12. Trap Building/Trap Creation
13. Climbing
14. Medicine
15. Astronomy
16. Hojo-jutsu (Art of Rope binding)

NINJA STYLE

The Ninja were not limited to these skills alone (they had much more skills than these) but these were the minimal amount of skills that they would have to learn in order to be operationally effective in the real time stress and fatigue of a real combat mission. The Ninja was known to possess many more skills than they revealed. The Ninja, having descended from Yamabushi or Mountain Ascetic Priests, lived a sober and quiet life that was full of study. They spent their time learning skills in a state of dedicated study and awareness. Their discipline made them able to learn at a rate that was incomparable with the Samurai class. The Samurai class spent their time plotting, practicing military warfare, and writing. The

NINJA STYLE

Ninja class spent their time learning and taking on any new skill that could allow them to defeat opponents with ease. The life of a Ninja was a quiet one that was full of practicing old skills, searching for new skills, and studying manuscripts and texts from which they could gain greater knowledge.

HAND TO HAND COMBAT

The Ninjitsu style of hand to hand combat is undoubtedly Jujitsu, Judo, and Karate techniques, performed in a Goju (hard and soft) manner. By blending together hard and soft techniques, the skilled practitioner of Ninjitsu made for a deadly warrior on the battlefield. Ninjitsu is the original Mixed Martial Art and because its form was honed over time based on

NINJA STYLE

experiences from real combat, it is not only effective but also deadly. Ninjitsu was made to be fast and able to end a confrontation in a manner of seconds. A ninja could not afford to waste time in physical confrontation, for example, during infiltration of a castle. A ninja had to swiftly dispatch one or sometimes two opponents and had to do so without drawing attention from other nearby guards. Ninjitsu depended on rapid disabling of guards if need be but the Ninja were not a blunt instrument of warfare like a Spearman or a heavily armed Samurai. The Ninja were quick on their feet and even faster with their techniques in taking out an enemy. The style that they used is known as Taijutsu. Though many of the Ninja

NINJA STYLE

may have specialized in various armed and unarmed styles, the main style that they used is referred to as Taijutsu. Taijutsu relied on being effective, being rapid, and being energy efficient. Punching and kicking would not kill a heavily armored castle guard. If hand to hand combat (unarmed) was to take place, the Ninja would seek to use the body weight and energy of his opponent against him by positioning himself in a way that would put his opponent at a disadvantage. Throwing a heavily weighted armored guard in a Judo like throw (which Taijutsu incorporates), would be devastating to an opponent, either instantly killing them or knocking them unconscious. The Taijutsu practitioner, in a

NINJA STYLE

demonstration that was not life threatening, would seem more like an Aikido practitioner. Aikido was influenced by Taijutsu because Morehei Ueshiba was himself at one time a student of Daito Ryu master Sokaku Takeda. The heavy use of deflection and evasive tactics made the Ninja a difficult opponent to counter and an even greater opponent to defeat. Their techniques were honed and mastered through rote mastery of throwing, blocking, choking techniques, bone breaking techniques, and striking. Their focus was on the perfect application of techniques combined with fast delivery of the movements. In order for the Ninja to have greater effectiveness, they most certainly relied on stealth. But there were and

NINJA STYLE

are many instances of Ninja having defeated Samurai in hand to hand combat, whether in an unarmed or armed situation. The style of Ninjutsu which the Ninja practiced incorporated the best elements of all the various Japanese and even Chinese martial arts, and this gave them an advantage when facing less skilled opponents on the battlefield. The Ninja were focused on principles and maintaining an adaptable and fluid stance in combat and this allowed them to defeat opponents that were even greater than themselves. The Ninja had to be adaptable and flexible on the battlefield because they were could and did face events that they could not previously plan for.

NINJA STYLE

OFFENSIVE WEAPONS

Short Sword – The Ninja would use
the Short Sword for killing an
opponent from close up and it was
also used a replacement in case the
Ninja lost his Long Sword.

Long Sword – The Ninja, being
originally Samurai, were experts in
the use of the Long Sword and were

NINJA STYLE

far superior to other Samurai in utilizing it in close range combat.

Ninja Throwing Star (Shuriken) – Shuriken were used as weapons of stealth to silence a castle guard and they were even used in hand to hand combat. Shuriken came in many shapes and forms, but their rapid velocity and swift delivery allowed Ninja to throw quite a number of them in a matter of a few seconds. In addition to throwing stars, the Ninja also used small lightweight throwing daggers, but these were more difficult to master than the throwing star. Some of the Shuriken were dipped in poison and some were explosive. They also threw Fire Shuriken or Ninja Throwing Stars that were on fire and would help to

NINJA STYLE

burn down a location. But their range allowed Ninja to take out an opponent from a range of 10 meters or more.

Bamboo Bow – Ninja were trained extensively in the use of the Bow & Arrow and were able to hit moving opponents from long distances with poison tipped arrows. Some of the Ninja had explosive arrows made from Bamboo that had a gunpowder cylinder inside of them. When lit, the

NINJA STYLE

Bamboo arrow would fire like a rocket at opponents.

Bamboo Staff – Ninja were master stick fighters that were extensively in the use of the 5-foot bamboo staff. The Ninja would spend much of their time undercover as a farmer carrying a staff and so the staff made a lightweight natural weapon that did not raise suspicion.

Curved Sickle – A deadly weapon in the hands of trained Ninja and was a weapon that posed as a farming tool which did not raise suspicion when walking with. They were usually carried in pairs and were a most deadly weapon in the hands of a skilled Ninjitsu practitioner.

NINJA STYLE

Blowgun – Ninjas used blowguns as mid-ranged weapons to shoot poison tipped arrows to incapacitate or kill castle guards. The blowgun had a double use in that it would also be used as underwater breathing apparatus during an escape.

Weighted Chain – Ninjas used the weighted chain or Manriki-Gusari in battle with high effectiveness. The Manriki Gusari was a heavy chain with two weights on other side of it. It was used for killing or even binding an opponent. The chain could be wrapped around the fist to make a shorter weapon.

Steel Claws – Ninjas wore steel claws or Tekko-Kagi on one hand or two hands, that could allow the Ninja

NINJA STYLE

to slash at an opponent or take away their sword without being injured. These steel claws resembled 5 animal claws on each hand and were most deadly in the hands of a skilled practitioner.

Dagger Ringed Rope – Ninjas used a special rope with a dagger on one end and a metal ring on the other end (which was used as a handle). This was used to catch and bind a castle guard from a mid-ranged distance.

Hand Grenades – Ninjas used specially designed hand grenades that were filled with gunpowder and fuse. Pieces of iron or even nails were fitted inside to act as shrapnel during the explosion.

NINJA STYLE

Tekken – The Ninjas had Knuckle Dusters or Brass Knuckles which they used during a hand to hand combat scenario. This may have been used as a last resort or as a weapon when within a town.

Improvised Weapons – The Ninja were able to use anything as a weapon and nature provided an excellent source for this task. The Ninja could spray a handful of dirt in an opponent's eyes to stun them before delivering the coup de grace. The Ninjas were masters of being able to innovate during a battle. Their use of improvised weapons ranged from natural elements like earth and rock to more engineered elements like projectile weapons.

NINJA STYLE

DEFENSIVE WEAPONS

Caltrops – The original defensive weapon of the Ninja came from rough fruit named Caltrops that gathered on river beds. The fruit was dried out until it was rough and because it contained jagged edges, pursuers feet would get ripped up stopping them in their tracks.
4 sided metal spikes that were thrown by Ninja as they were running to stop pursuers.

Smoke Explosives – Explosives that were used to emit smoke, rather than to destruct. These were thrown to mask the location of the Ninja and to confuse pursuers during an escape.

NINJA STYLE

NINJA TOOLS

Grappling Hook – The Grappling Hook is a long rope with a hook at the end of it, that the Ninja used to mount a castle wall or climb a building. The Grappling Hook gave a superior advantage when scaling castle walls. The Grappling Hook was primarily used by Ninja during infiltration though it could have also been used for escape. The Ninja had extensive training and expertise in the use of the Grappling Hook. It was a main Item for infiltration and it was carried on missions regularly because of its capability. The hook was most likely comprised of iron and the rope was probably made of Hemp fiber, making it durable and long lasting.

NINJA STYLE

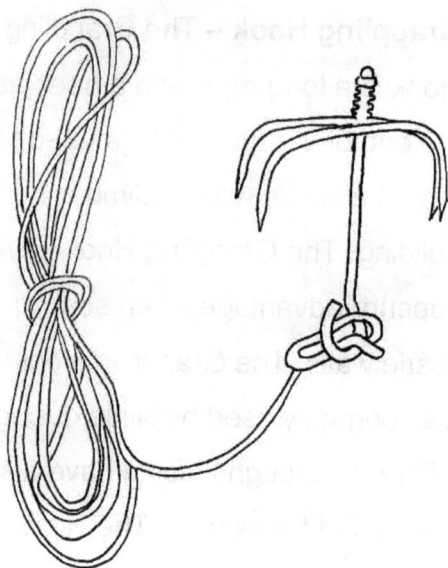

Spiked Hand Claw (Shuko) – Metal plate that slipped over the hand and featured 4 sharp spikes on the palm. Was used for climbing walls and scaling buildings, but could also be used as a weapon. Shuko gave the Ninja the ability to perform

NINJA STYLE

extraordinary feats of climbing with ease.

Pick – Ninja carried a long sharp pick which they used to drill holes in walls that were to be climbed and to drill holes in walls for eavesdropping.

Climbing Ladder – Ninjas used a climbing ladder or Shinobi Kumade for scaling the walls of a castle. The Shinobi Kumade was made from a rope pushed through pieces of cylindrical Bamboo wood with a hook attached to one side of the rope.

Kunai – The Kunai was a mix between a dagger and a tool. It could be used for drilling a hole in a wall or for dispatching an enemy at close range. It is a distinctively Ninja tool

NINJA STYLE

whose use was only taught in Ninja schools.

NINJA CLOTHING

The Shinobi Shozoku is the Ninja uniform. As the Ninja were for the most part of their mission in disguise, the only times they would be wearing such a uniform is during actual operation of infiltration and escape. Some historians argue that the Shinobi Shozoku came from the Japanese theater and was originally used by individuals that were puppeteering. The dark shade of the uniform would blend in to the dark shadows of the theater so that the spectators could wholly concentrate on the actual puppets that were moving. Some modern scholars argue that the Ninja had no official

NINJA STYLE

uniform but it is more likely than not they wore a uniform, if not identical, at least similar to the Shinobi Shozoku. They may have worn Kusari or Japanese chain mail armor during surveillance, but it is more than likely that they wore no Kusari doing their silent operations as Kusari would have made so much noise that the Ninja's position would be given away. In addition, any Kusari would have slowed down the Ninja during infiltration or more importantly during an escape. The Ninja clothing, because of its dark hue, would not give off any shadow that would betray the position of the Ninja. In addition, because of its dark hue, the Ninja was able to easily hide in shadows located between walls and buildings. Its lightweight

NINJA STYLE

material made it ideal for movement and maneuvering under the intense pressure of Ninja operations. It is debated if indeed the Ninja wore pure Black uniforms or if they were a dark Brown or dark Red. They wore dark uniforms whether the colors were Black or dark Red/Brown. As it was the uniform favored by puppeteers in the theater to hide themselves during a performance, the color had to be dark and not reflective of light. Any material with a color that could reflect light would betray the position of a Ninja, making them open to danger. Regardless of the color they chose during their mission, the mission did indeed wear a dark colored uniform that would hide their physical presence and negate any light source.

NINJA STYLE

KOGA-RYU

The Koga-Ryu is one of the most famous Ninja schools in Japan besides the Iga-Ryu. The Koga-Ryu school of Ninjitsu was founded by 53 families (clans). The birth of the Koga-Ryu is said to have come about when the study of Illusion (Genjutsu) met the study of Guerrilla Warfare. Daimyo Takeda Shingen used the services of the Koga Ninja extensively and relied on their ability to assassinate enemies un-reachable by conventional warfare. The Koga-Ryu pioneered the use of

NINJA STYLE

The Ninja faded out as a force in Japan after 1603 C.E.

NINJA STYLE

explosives and firearms for Ninjas. Ninjas had a small palm pistol which they used to deliver a deadly single shot round at close range. Koga-Ryu pioneered the use of smoke grenades to disappear in to thin air. The Koga-Ryu's training in illusion and guerrilla warfare, as well as other vital sciences like medicine and astronomy, made them second to none in the arts of the Ninja. The Koga-Ryu had advanced understanding of the principles of illusion and this enabled them to pull of great feats that made them famous throughout Japan. Many of the Iga Ninja came from the Koga Ryu and the similarities between the Koga-Ryu and Iga-Ryu are readily apparent in their respective styles.

NINJA STYLE

KUNOICHI

Kunoichi were female Ninjas that were highly skilled, and better able, in many instances to carry out missions that male Ninja were not able to do. Kunoichi were faster in climbing, faster in movement, and more silent than their male counterparts. The greatest (and probably the most famous) female Ninja was Mochizuke Chiyomi and it is believed that she was a direct descendant of Mochizuki Izumo. Chiyomi built an All-Female Ninja school in Koga and trained between 200 to 300 Ninjas. Female Ninjas used various disguises in order to travel through Japan un-detected and they include:

NINJA STYLE

a. Theater Actress – Many female Ninja posed at professional entertainers.

b. Prostitutes – Many female Ninja chose this disguise to move about it knowing that it would not raise suspicion.

c. Religious Pilgrims – This was an innocent disguise for female Ninja and this allowed them access to key leaders.

d. Priestess – This was the main disguise used by the female Ninja of Koga.

Kunoichi, like their male counterparts, were not blunt instruments of warfare. Although the

NINJA STYLE

Kunoichi were deadly in battle, Kunoichi were mostly used to infiltrate enemy areas and to gather information. The Kunoichi use of disguises made them able to enter every building and being female made them not worthy of suspect. A alms begging female pilgrim would raise little or no suspicion, as women were viewed as being frail and not able to physically contend with a man. The Sengoku Jedai period in Japan proved otherwise. Not only were female Ninjas as efficient and effective as their male counterparts, but in many instances they were more effective because of their ability to make an opponent drop his guard down.

NINJA STYLE

TOIRI NO JUTSU

The Ninja were masters of Toiri No Jutsu or the Art of Infiltration. They were able to use multiple techniques to enter in to an enemy's castle in complete silence. There were certain conditions that the Ninja would look for before beginning their infiltration in to the enemy's castle. First, they would wait until the time of the month which the Moon light was the least bright. They would have conducted surveillance beforehand to know the exact times that the guards are changed and where the locations of the guards are. Ninja were master tacticians as well as strategists. All of their time in preparation allowed them to execute perfect operations with little or no mistakes in the process. Months of meticulous

NINJA STYLE

planning would be done beforehand in order to create one opportunity that would last less than one minute. It was this level of work ethic and dedication to perfection that allowed them to gain the reputation of being demon like and more than human. The exploits of the Ninja were personified as being supernatural because of their high success rate in infiltrating an opponent's heavily guarded castle. Being that the Ninja were masters of infiltration, they would sometimes use multiple methods and various steps in sequence in order to create an opportunity for infiltration. The object of infiltration had to be silent and swift as well as being traceless. Castle guards usually dispatched Ninja hunter patrols to search Castle

NINJA STYLE

grounds for hiding assassins. Certain steps were usually taken before a Ninja would begin their infiltration and these were done to make sure that the operation would be a success.

Pre-Infiltration Steps:
1. Know the Layout/Roads

2. Draw a Map of the territory

3. Devise a Strategy for Entry

4. Rest Well

5. Pick a Night Where the Moonlight is least

6. Pick a time that will be the quietest.

7. Assemble items needed for mission.

NINJA STYLE

Infiltration Techniques

1. Enter in disguise – The easiest way to infiltrate is to pretend to be worker. This was used extensively by Ninja to infiltrate without raising suspicion.

2. Enter by Employment – The Ninja would be hired by the enemy, not know that the Ninja was indeed a Ninja.

3. Enter using Fire Diversion – The Ninja would create a fire diversion in order to create chaos, allowing them the time and opportunity to enter unnoticed. One of the best diversion used by the Ninja was fire.

NINJA STYLE

4. Enter using Water – Water can be used to flood an area, creating an opportunity to enter.

5. Enter by Climbing – The Ninja would climb walls and scale buildings in stealth in order to reach the inside of an enemy castle.

6. Enter in Deception – Pretend to be a part of the Castle workers and enter.

7. Enter with Enemy – Ninjas would capture an enemy soldier and force them to bring the Ninjas inside quietly.

NINJA STYLE

8. Enter with Confusion – Ninjas would create noise confusion and confusion by smoke, in order to create an opening to enter a castle during chaos.

9. Enter by Ambush – Ninjas would draw out the enemy from a castle and ambush them in order to enter a castle.

10. Enter by Betrayal – Get a person in the castle to betray their liege and bring you in.

11. Enter by Boring – Ninjas would use special Picks to create holes in walls to enter through as well tools to break/pick locks.

NINJA STYLE

Entering a castle by moving silently past all the guards was the first step in a Ninja's operation. Once inside, they had various tasks they had to complete and this would have included things like doing sabotage, stealing a valuable item, rescuing a hostage, or setting a castle to fire. The infiltration of Ninja would, in many cases, be aided a nearby military force that was allied or a part of the force of the Daimyo that had hired the Ninja. The military force would use a form of attack to create a diversion on the enemy's castle before a Ninja would go through the arduous task of entering an enemy castle without being spotted. The attack could be in the form of arrow raining down on castle defenders or in the form of a fire to create

NINJA STYLE

confusion. The use of a military force to creating a gap by surprise for the Ninja to enter was both intelligent and efficient use of forces and was successful in application. The Ninja were not reserved to one technique for infiltration and would use the most realistic form of entry.

CLIMBING

The skill of climbing was an extremely vital skill to possess for their occupation. Climbing was seen as a sport or game by Ninja, and they used to frequently gather together to see who could climb the farthest and highest. The Ninja developed many tools to assist them in climbing and they include, grappling hook, climbing hooks, and wooden rope ladder. In addition, they

NINJA STYLE

had the wearable spiked palm bands they used to free climb up walls. The Ninja viewed climbing as one of their main skill sets and without it a Ninja was wholly unprepared to meet the challenges that came with infiltration. Ninjas would use disguises and guile to enter an enemy's castle in broad daylight, but this option was only available when it presented itself. Ninjas did not always have this option so in order to infiltrate they would have to climb, sometimes one or more walls, while contending with castle guards and the noise that movement creates. In addition, most castles had patrols that would search for Ninjas without a torch to give off their location. The Ninja would have to be an expert in the climbing of walls to complete the mission.

NINJA STYLE

ESCAPE TECHNIQUES

Escaping a location is the opposite of infiltration of a location. Certain steps have to be taken and they include:

1. Study the times and patterns of movement among guards

2. Be physically and mentally prepared to leave. Leaving may take a greater mental preparation than entering so you should dedicate time to positive self-reinforcement and think about the good things that are waiting for you.

3. Pick a time that is quietest, usually before the suns come up in the morning.

NINJA STYLE

4. Pick a night that has the least amount of moonlight, usually between 14 days to 21 days in to a month.

5. Have a diversion prepared in order to misdirect guards if movement is heavy.

6. Have places picked out in the line of escape that could be used to hide you; Physical objects.

7. Have a weapon (or makeshift weapon) with you to defend yourself during the escape. The point is not to fight out, which is an impossibility. If you are seen, then the

NINJA STYLE

probability that you will be captured is 100 percent.

Just like infiltration, the Ninja had only chance during escape. They would have to go to great lengths to ensure that they would be meticulous in their planning to prevent anything from surprising them. Their escape from an enemy castle would have to be as quiet and non-alarming as their entrance would be. They could not break their way out by fighting so they had to sneak their way about using silent and undetected movement. During an escape if caught, the Ninja would most likely be tortured and executed. Being detected was not an option, which is why Ninja were extremely

NINJA STYLE

careful in escaping, using even more discretion than during infiltration.

INVISIBILITY

Of course the Ninja were not able to actually disappear, but to the human eye it did appear so. The Ninja would use natural principles to blend in with their environment, giving the illusion of disappearance.

Air – The Ninja would climb up to areas that they could not been seen in like trees, rooftops, and walls. They would use height to avoid being observed by passing castle guards.

Water – The Ninja would use their blowgun as an underwater breathing

NINJA STYLE

apparatus to hide in water undetected.

Fire – The Ninja would only move behind light sources in order to prevent their shadow from being generated. Movement in shadow and avoiding lighted areas kept Ninjas undetected even in a heavily guarded castle.

Earth – The Ninja would use small shrubs, dirt, and leaves to cover their bodies when hiding during an escape, making them seemingly invisible.

Houses of Ninja Clans and Ninja leaders were constructed in a manner that would allow for the rapid disappearance of a Ninja during a

NINJA STYLE

raid on the house. Trap doors, tunnels, and ingeniously created hiding places were injected in to the plans of the house that was being constructed. In many instances, Samurai would raid in to a Ninja leader's house, only to find that he had disappeared. Donden-gaeshi were special doors that were placed in walls. When pressure was put on the Donden-gaeshi (the revolving door), the Ninja would be able to disappear in to the wall instantly. Tunnels were built under houses by Ninja to be able to escape quickly. In each of the rooms of a castle built or designed by Ninja, there hidden devices that would be used in case if the castle or residence they were occupying came under an attack. This also gave the Ninja the

NINJA STYLE

advantage of being forewarned in order to not be surprised during an ambush on their home. The Ninja would use such devices as building a ladder from inside a cupboard leading to another floor, which would allow them to escape silently and without a trace. The used of hidden passages between floorboards gave them a distinct advantage in hiding and disappearing. There is one particular instance in that Samurai raided a Ninja's castle only to find that the Ninja had disappeared completely and this was done by hiding inside of a chest full of Buddhist mantras. The ability to disappear at will gave the Ninja a reputation of being supernatural powers and being wizard-like. But their ability to disappear at will did

NINJA STYLE

not come from magic spells. The
Ninjas ability to become invisible
instantly was because of their trained
use of smoke grenades and other
illusion enabling devices that allowed
them to create the illusion of
invisibility. The Koga-Ryu were
highly skilled in the use and study of
Illusion (Genjutsu) and this enabled
them to do things that would make
not only Samurai but other Ninja
confused. The use of illusions played
a key role in the art of Ninjitsu and
without it, the Ninja would not be
Ninja. The birth of Ninjitsu (arguably)
came about when Guerrilla Warfare
met the art of Illusion. Con-Artistry &
lying are in themselves illusions, that
use words to manipulate the minds
of others and the environment.
Shinobi would be sent behind enemy

NINJA STYLE

lines with the express instructions to bring out some type of change in that area. The Shinobi were able to used advanced psychology based techniques to manipulate public opinion and they were so effective that they were able to bring out political upheavals and revolutions in a given area. But their subtlety was their key advantage because any operations they undertook to change or manipulate public opinion occurred slowly through a process that made the change seem natural and locally developed. It had to appear so or else the individuals involved in bringing about the upheaval would have been arrested and executed.

NINJA STYLE

CON-ARTISTRY AND LYING

The Ninja had to spend long periods of time in complete or partial isolation, sometimes going for months without speaking to anybody. This increased their ability to sense their surroundings to a greater level than their enemies. Sensory deprivation, strengthens the senses instead of weakening them. But when they did speak, they could not reveal really who they were or else it would end their mission and possible even end their life. The Ninja were trained in the ability of lying and knew how to lie in a manner that would make fact checking them impossible. They would spread lies or misinformation often in a province in order to create things like political change or an uprising. Their use of

NINJA STYLE

con-artistry was also well known as was their use of cold reading techniques. As master manipulators, the Ninja had to be able to bring about change within an area in a very short time on behalf of their liege. Their use of cold reading techniques to manipulate individuals in power and their use of illusion gave them the ability to gather or to destroy public opinion in a relatively short period of time. Shinobi is just another named used for Ninja, depending on what area of Japan you were living in. But whether they were referred to as Ninja or Shinobi, their ability to influence large amounts of people without overtly doing so, made them not only effective with weaponry but also equally effective via the use of

NINJA STYLE

language. The Ninja were simply masters of the principles of psychology hundreds of years before their existed a science by this name. Illusion works on the principles of the science of illusion as well as incorporating understanding of the science of optics (line of sight). The Ninja's ability to intelligently use the principles of psychology against their opponents made them feared by their enemies and allies. But this superior training also gave the Ninja a distinct advantage and this made the Ninja highly sought after by feudal Warlords (Daimyo's) that sought out quick and painless ways of disposing of their enemies.

NINJA STYLE

ILLUSIONS

The Ninja used many disguises to be able to infiltrate enemy castles, but no disguise was greater for them than that of a peasant looking for work. Many of the Ninja were Warrior Monks and were heavily trained in the use of spirituality and Buddhist mantras. Some historians attribute the Ninja's invisibility techniques as reason for being able to blend in but in reality is was their superior skill set in the creation of disguises that allowed Ninjas to make silent entries. Ninja wore straw hats to block their faces being when moving. They would carry farming implements to appear to be a farmer or they would carry fishing nets to give off the impression of being a fisherman. Because the Ninja had to spend long

NINJA STYLE

periods of time behind enemy lines, they would usually also be skilled in one or sometimes two professions in order to blend in entirely with the population. The various professions that the Ninja used included:

1. Merchants – The Ninja would engage in the profession of being travelling merchants, which would also give them access to local officials.

2. Theater actor – The Ninja would work in "Noh" (theater) plays and this gave them access to popular society as well as gave them first-hand experience in the creation of elaborate disguises.

NINJA STYLE

3. Mendicant (Beggar) – The
    Ninja would play the role of a
    beggar monk, looking for
    money to continue his spiritual
    quest.

4. Farmer – The Ninja would
    play the role of a Farmer and
    carried around farming
    implements in order to further
    bolster their claim. Their
    disguise was not skin deep
    however as their superior
    knowledge of farming made
    them seem legitimate.

5. Fisherman – The Ninja would
    play the role of a Fisherman
    and they would carry around a
    fishing net with them in this
    role. As Japan is an island

NINJA STYLE

nation, the role of a
Fisherman was not only
natural but not questioned.

The Ninjas were masters of disguise
creation and they used this to their
advantage to penetrate heavily
guarded enemy castles without
raising an alarm. Their uses of
disguise were done in a very natural
way. A Ninja would become a
beggar looking for spiritual alms. A
Ninja could become a merchant,
selling his goods and raising funds
for his next mission simultaneously.
The theater in Japan represented
Pop Culture in Japan and theater
actors and actresses were no less
revered by popular culture, than they
are today. The theaters also gave
Ninja access to rumors.

NINJA STYLE

The Ninja were highly trained in the use of illusions, both psychological and actual. Their skill set in illusion planning, illusion creation, and illusion execution gave them a superior advantage when facing a heavily guarded castle full of well-trained samurai. The Ninja's use of psychological illusions included the creation of rumors that would help to give them an advantage in a situation. For example, if the Ninja were living near the castle of a Daimyo that they sought to assassinate, then they would spread a rumor that an attack was imminent by an enemy, causing the Daimyo to gather villagers nearby in to the castle for military recruitment. Thus, the Ninja would slip in un-detected in to an enemy's castle while in plain

NINJA STYLE

sight the entire time. The Ninja's use of illusions also included using them in the construction of a house, allowing them ample places to not only hide but also to escape. Secret doors leading to secret passageways or tunnels gave them the ability to disappear at will. The use of smoke grenades to create a mist thick enough to disappear in to also gave them the reputation of being able to disappear instantly. The Ninja were also trained in illusions that would help to create diversions. The use of diversions were an adaptable and real time illusion that had to be performed during an escape or infiltration. Illusions ranged from the simple to complex. Much of the warfare that was conducted, whether conventional or guerrilla warfare,

NINJA STYLE

was based around the principle of deception. By using illusions to create traps for their enemies or by using illusions to escape from a castle full of enemies, the Ninja took on a reputation of having super-natural powers. The Ninja had and have no super-natural powers and all rumors of the Ninja having them can be passed off as myth, legend, and folklore. The Ninja had and have advanced understanding of the use of illusions. This is why they were always at an advantage to their enemies and this is why their enemies feared them so greatly. The Ninja could appear and disappear from anywhere given just a very short time. This made them despised by enemies and loved by allies.

NINJA STYLE

MEDITATION

The Ninja believed in being able to harness supernatural like powers through the use of Buddhist hand symbols (Kuji). They believed that by performing the hand symbols in a certain order, they would be able to harness certain powers of nature. Their deep connection with nature may have been the primary reason that their system depended on the harnessing of the forces of nature. The majority of the Ninja came from Yamabushi or Warrior Monks and this carried over in to their Ninja lifestyle through the implementation of Buddhist hand symbols as a form of gaining the advantage on the field

NINJA STYLE

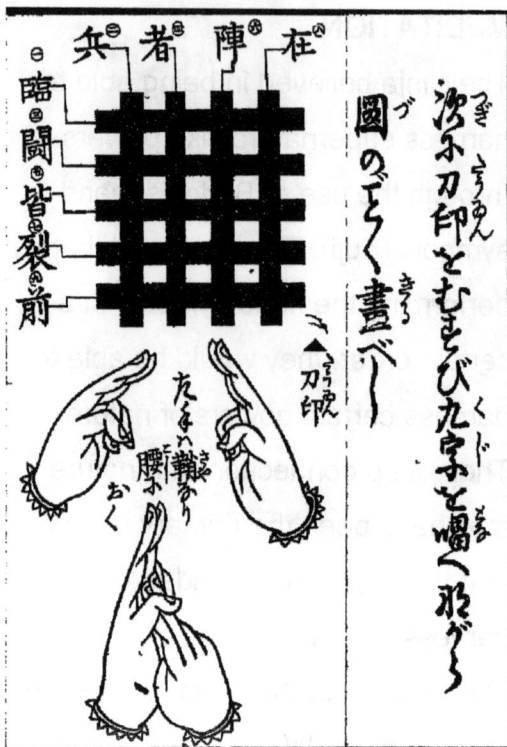

of battle. The sequential order in
which the Ninja performed the
Buddhist hand symbols is a secret
(Densho) that has been passed
down orally by generations of Ninja.
The Ninja would perform the hand
symbols in a certain order to be able

NINJA STYLE

to create a supposed effect, such as invisibility. It is more likely than not that these hand symbols were performed as an end all to themselves as a form of ritual. More likely than not, these hand symbols had no effect on the performance of the Ninja other than to bolster them with a sense of relaxation and tranquility before they carried out their task. In the West, the performance of these various hand symbols has presented it as being a form of "Ninja Magic". A practitioner of Kuji Kiri was in fact referred to as a majutsushi which meant a conjurer. Kuji Kiri or the "Nine Symbolic Cuts" were the 9 hand symbols used by the Ninja. The practice of Kuji Kuri is a part of the Buddhist esoteric tradition of the

NINJA STYLE

Ninja that was derived from Taoism from China. It was used as a form of meditation in order to prepare the mind before the Ninja started their mission. First a Kanji symbol was drawn on paper. Then 9 lines were drawn in a horizontal format over the symbol of desire. A 10th cut was symbolically drawn (without ink) from the top right hand of the grid to the bottom left part of the grid to symbolize the start. The 10th and final "cutting" motion was supposed to make the magical grid become active. Kyo, Sha, Jin, and Zai are located on the vertical lines and Rin, Toh, Kai, Retsu, and Zen are located on the horizontal lines.

NINJA STYLE

	KYO	SHA	JIN	ZAI
RIN				
TOH				
KAI				
RETSU				
ZEN				

4 Lines were drawn vertically and 5 lines were drawn horizontally. This created a power grid similar to a Talisman. This itself can be viewed as a form of Talisman that is attempting to harness the forces of nature to achieve an end goal. Based on this definition alone for magic, the activities of the Ninja were viewed as conjuring.

NINJA STYLE

The 9 Kuji Kiri Hand Symbols

Rin – Strengthening mind & body

Kyo – Focusing and directing energy.

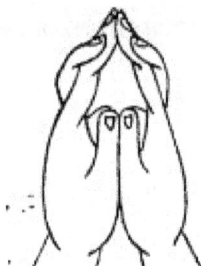

Toh – Harmony with environment.

NINJA STYLE

Sha – Healing power over yourself and others.

Kai - Sense danger & increase senses. To feel what other's feel.

Jin - Knowing the thoughts of others.

NINJA STYLE

Retsu – Mastering time and space.

Zai – Control over the elements of nature. Creation.

Zen - Enlightenment

NINJA STYLE

MEDICINE

Ninjas were skilled in and contained expertise in the use of various Medicines. They knew which plant could help you with high blood pressure and which plant to use if you had diabetes. They had advanced understanding of the use of herbs and natural medicines. They acted as their own bonesetters (Doctors) and they also administered their medicines to non-Ninjas as well. They would brew up their own medicines and concoctions, and they created informal recipes for the healing of patients. Ninjas were highly trained in the use of herbal medicines and chemicals, possessing the answers to medical problems that Western physicians

NINJA STYLE

would only discover hundreds of

years later. Their recipes ranged
from normal herbal medicine like
picking leaves from a plant and
brewing them to very complex
recipes that included animal fluids
and other items that incorporated

NINJA STYLE

NINJA STYLE

Shinto philosophy. The Ninja's connection with nature and worship of Kwannon (Mother Nature), made the appreciation of nature a priority for them. Kwannon was not viewed as a deity in the Western sense of the word but rather was viewed as a Saint. Kwannon or Kannon, had many temples in Japan related to her but she was sometimes shown in an androgynous manner. It has been rumored that the original name for the company Canon originated from Kwannon or Kannon, and a press release on the Canon website celebrating the 80th anniversary of their "Kwanon" camera confirms this fact. The Canon website states that "The engineers who created the camera decided to name it after Kwannon, the Buddhist deitydess of

NINJA STYLE

mercy, hoping the deity would share her benevolence as they pursued their dream to produce the world's finest camera. The camera's lens, called Kasyapa, after Mahakasyapa, a disciple of Buddha—also took its name from Buddhism. Additionally, the top portion of the camera body featured an engraving depicting the thousand-armed Kwannon."

Source: Canon.com website

NINJA STYLE

Kwannon, also known as Guan Yin, came to Japan via China and was worshipped as a male deity before the 12th century. Kwannon was a Taoist deity that was imported in to Japan and took on a female form or even androgynous form. Taoists view Guanyin (Kwannon) as a male deity but it is important to note that his transformation in to a female one came about under the patriarchal Japanese culture. If the Japanese was and still is so patriarchal, how is that the most important deity in Japan was a female that was worshipped by men as well? Kwannon popularity was widespread throughout Japan and her temples were regarded as mandatory places to be visited. The belief in Kwannon was widely practiced in Japan and

NINJA STYLE

pilgrimages to Kwannon temples were widespread. Invocations to Kwannon were said to break the chains holding individuals back from progress and thus Kwannon was seen as a Creative saint rather than as a Destructive one. The Shinto belief in harmony with nature also influenced the Ninja greatly and thus being healed by the fruits and plants of nature was for them, in a way, a re-connection with the source of all life. Medicine played an important role in the life and even operations of a Ninja. Medicine provided the medium for which nature could heal them.

NINJA STYLE

ASTRONOMY

Ninjas had excellent understanding and applicable knowledge of Astronomy. Their knowledge of Astronomy would allow them to know what part of the month had the least Moon light and which part had the most Moon light. Astronomy allowed Ninjas to travel without any type of direction finder for months at a time just by using the stars. The Ninjas used stars for navigation, whether on land or on water, and it was this knowledge that allowed them to move through the countryside of Japan so easily. The Ninjas were trained and gained at least a workable experience, if not expertise, in the use of Astronomy to aid them on their lengthy missions.

NINJA STYLE

HOJO-JUTSU

The Ninja had specialized training in Hojo-jutsu or the Art of Rope Binding. The origins of this ancient art form are argued far and wide but it has been impossible to state who is in fact its originator. What is known with certainty is that this is a distinctly Japanese art with no counterpart in any other nation on earth. The ropes that the Ninja used were most likely created from Hemp. The intricate patterns for binding that the Ninja used made it not only effective, but visually (aesthetically) pleasing as well. The Ninja were experts in this art and received sufficient training in it as it was a vital part of their skill set. The Ninja were able to, within a matter of seconds, bind up the hands and feet of castle

NINJA STYLE

guard or an opponent they wished to capture. Hojo-jutsu has various techniques for binding up one or more individuals use just rope and ingenuity. In many instances, Ninja had to capture (alive) an enemy and deliver him to an opponent. Hojo-jutsu gave the Ninja a tactical advantage in controlling a situation relatively fast, without having to resort to killing. The Ninja, being trained in not only Medicine but also the Art of Rope Binding, had an advanced understanding on the human anatomy which allowed them to use Hojo-jutsu for not only binding, but they would also be used for torturing or even killing an enemy. The knowledge of Hojo-jutsu gave the Ninjas a definite advantage in the execution of operations.

NINJA STYLE

GETTING CLOSE

The Ninja would learn so many different skills and practice them constantly in order to have the advantage when facing an enemy or while completing a surveillance mission. The most important mission of all was getting close to the enemy. The Ninja could not be effective from a distance and the positioning of the Ninja close to the opposition was vital for the mission to be completed. Getting close to the enemy could be considered the first part of any mission and without achieving this vital step, the rest of the steps involved in the mission like surveillance, misinformation spreading, sabotage, and assassination would have been impossible. Getting close to the Ninja

NINJA STYLE

was the first and most important step of the mission. The Ninja, as discussed previously, used multiple costumes and disguised to achieve their goal of getting closer. Although getting closer to an enemy seemed trivial in comparison to the mission, if done incorrectly could result in the death of the Ninja as well as betrayal of the plans the Ninja held. Getting closer to the enemy was different than infiltration, which required its own methods for achieving the Ninja entering an enemy's caste. Getting closer would entail preparation that involved one or multiple reconnaissance missions to gather enough information in order to decide what approach should be taken for the Ninja to in fact move closer in to a position that could

NINJA STYLE

potentially allow for infiltration in the future.

LOCK PICKING

In many instance, Ninjas would be contracted to steal an item or to place an item in a location that could disrupt his enemy. The Ninjas were masters of picking locks and had advanced training in the dismantling of complex locks. The Ninjas had manuals that laid out techniques as well as Densho (secret teaching) about the various methods of opening wooden doors with iron locks. This knowledge was vital to their missions and without it such advanced first-hand knowledge of the workings of lock, it would have been impossible (in many missions) to enter areas that had been locked

NINJA STYLE

down. The Ninjas used techniques ranging from opening locks without noise to breaking a lock to open it. The Ninjas carried lock picking tools with them to complete their task and these were iron or steel precursors to the modern lock picking sets used by burglars. The Ninja could consider lock picking skills as being a key skill they would have to learn in order to complete their training. In addition, the Ninja knew how to create locks because they understood the principles of how locking mechanisms work, giving them again a huge advantage on their missions. It was their ability to combine all the various skills together and use them in unison to complete their missions that made the Ninja incomparably efficient.

NINJA STYLE

NINJAS AS STRATEGISTS

The Ninjas were thought of as individuals that were solely sent to carry out vital missions for various Daimyo (feudal landowners). In reality, the Ninjas were able strategists that used their tactical and operational knowledge to act as advisors for Daimyo. The Ninjas were literate and well versed in various strategists' methods. The Ninjas read and memorized Sun Tzu, applying his minor and major strategies to issues that the Daimyo were dealing with. The Ninjas vast applicable knowledge of military operations made them capable advisors and their strategies gave their paymasters a huge advantage over Daimyo that did not hire Ninja advisors. The first thing that an

NINJA STYLE

informed Daimyo would do is hire a Ninja advisor to give them information about the area and their enemies. The Ninja clans held information about all the various Daimyo as well as about their vassals. This information was vital to the Ninja's work and allowed them to have the specialization to advise the Daimyos. The depended on this information and were unable to be so effective in engaging their enemies without this information. The advice that the Ninja delivered to Daimyos was based on their vast surveillance networks that spanned not only large provinces, but all of Japan. The Daimyo did engage in many battles without the use of Ninja, but doing so did not make them stronger. The Ninja brought an advantage to

NINJA STYLE

Daimyo and their Samurai. The use of Ninjas inside and outside of a castle with a signaling system, allowed for the Daimyo to know where to start an attack from. In some cases, the infiltration of Ninja in to a castle allowed for the taking of an entire castle with minimal damage caused to it and with a minimal amount of human casualties. Samurai battles were hard fought, bloody, and ultra-violent. Samurai were called out to take over a province, lay siege to an enemy castle, or to destroy an opposing army. The infiltration tactics of Ninja allowed a Daimyo to defeat an opposing army by killing their Commander and saving potentially tens of thousands of lives by preventing Samurai having to slice

NINJA STYLE

and hack each other in to bits on the battlefield.

NINJA MORALS

The Ninja were highly moralistic and they refrained from the drinking of alcohol. They believed that alcohol was the main reason that a Ninja would reveal information. The Ninja, rightfully so, understood that alcohol made Ninja careless and talkative, inducing them to reveal information that could lead to their own death as well as betraying the mission they had been assigned. The Ninjas were also asked to refrain from lustful behavior that could potentially lead to them betraying their mission. The Ninjas also abhorred greed and greedy behavior, seeking a natural and simple lifestyle instead. As many

NINJA STYLE

of the Ninjas came from a tradition of Yamabushi or Ascetic Mountain Priests, the simple lifestyle was both understandable and comfortable for them. The pursuit of material riches and wealth was viewed as innately weak in character and a sign of a troubled spirit. The Ascetic Mountain Priest tradition of Shugendo greatly influenced Ninja thought and philosophy which promoted harmony with the environment and harmony with people. The Ninja were well read and intellectually sophisticated people that exercised control over their character and sought to gain progress in becoming better, not only skill-wise, but morally and ethically. The development of character was a part of Ninja life and they honed their characters by living lives of

NINJA STYLE

simplicity. The Ninja are viewed in popular culture as being almost horror movie like in their adherence to death and destruction. It is just the opposite. The Ninja sought harmony with their environment and it was widely believed by Ninja that the use of its skills to promote self-interest would result in doom. The Ninja carried out missions based on contracts by Daimyo and their Samurai vassals, but it was up to the Ninja clan entirely to choose a mission or to deny a Daimyo. The Ninja clans were independent and thus they chose which Daimyo to work with and for. Coming from a tradition of Shugendo which favored harmony, the Ninja created greater harmony by choosing Daimyo that could greater harmony. During the

NINJA STYLE

Sengoku Jedai period, harmony was achieved either through diplomacy or it was achieved on the battlefield. The Ninja provided an opportunity for Japan to achieve harmony in a smoother fashion than Samurai killing each other in large numbers on the battlefield. The Ninja were the 5th wheel in the equation, but a vital one. There were smaller Ninja organizations that went rogue and used the arts of the Ninja to solely carry out missions based on money. But for the most part, the Ninja maintained some form of loyalty to Daimyo clans that they perceived to be able to unify Japan and to prevent further warfare. The Ninja did not take part in competing with the Daimyo, nor did the Ninja take part in creating greater confusion and war

NINJA STYLE

during the Sengoku Jedai period.
Rather it was the Ninja that were
more loyal than Samurai during this
period and it was the Ninja that
provided the key skills need to
prevent further bloodshed on the
field of battle. Just like how Alfred
Nobel invented TNT to prevent
further war or how Samuel Colt
invented the revolver in 1836, Japan
invented the Ninja to rid itself of war
all the faster. The negative image
that is portrayed of the Ninja as
being the evil in Japan has been
created by popular culture and the
movies being churned out of
Hollywood. The Ninja, despite the
negative image that is painted on
them, were not only highly ethical
and moralistic, but also highly loyal

NINJA STYLE

at a time when Samurai changed
loyalties on a weekly basis.

NOT LONE WOLVES

Despite the popular myth that Ninja
worked alone, there is evidence that
shows that higher ranking Ninja
would make use of lower ranking
Ninja to assist them on their
missions. Whether it was
surveillance, intelligence gathering,
or an assassination, a lower ranking
Ninja would provide assistance to
the higher ranking Ninja so that their
mission would become a success.
This would also give lower ranking
Ninja the opportunity to gain real
time experience in the field. The
lower ranking Ninja would participate
in the operation of the actual mission
but would serve as a helper in the

NINJA STYLE

carrying out of various task related to the mission. The lower ranking Ninja would in some cases travel with the higher ranking Ninja, providing not only protection but also logistical assistance. In some cases, Ninja would have to setup a temporary base near the area that would be infiltrated and the lower ranking Ninja allowed the higher ranking Ninja to concentrate on the more important matters of the mission while the lower ranking Ninja could see to the infiltration/escape base. The use of lower ranking Ninja allowed for them to learn vital knowledge that could not be attained in a training scenario. The lower ranking Ninja were also proud to be able to gain the honor and fame that came with

NINJA STYLE

participating in the operation of a successful mission.

NINJA RELIGION

Confucianism, Taoism, and Buddhism played a major role in the life of Ninja. Ninja were taught that in order to have Divine protection, they would have to be religious. The Ninja, having descended from the Shugendo Ascetic Mountain Priest tradition, were highly religious. Their adherence to harmony with nature and their environment made them highly spiritual. Lecturer Milton Terry stated that "a deeper and more widespread influence than that of anything of Chinese origin was the introduction into Japan of Buddhism, which was first brought in 552 AD, but did not succeed in leavening the

NINJA STYLE

whole country until the middle of the ninth century. It was quietly propagated by leaders of various Buddhist sects which differ in minor practices, and slowly it gained ascendency, but its first more notable triumph followed the teaching of Kukai, founder of the Shingon sect, who so adapted Buddhist doctrines to the traditional ancestor worship as to maintain that all the Shinto deities were incarnations of Buddha. With great ingenuity and cunning, a new interpretation was given to ancient myths, and new constructions were put upon old beliefs. The Shinto deitys, rites, customs, and traditions took 011 a Buddhist significance, and many of the mysteries of birth and of death were explained in a

NINJA STYLE

manner so simple and popular as to commend them to all who listened to the new teaching. For Buddhism had already learned in India and in China the clever art of appropriating old beliefs and customs and of clothing tin-in with a new and higher meaning. Confucianism itself had already in part prepared the way for Buddhism in Japan, and the successful Buddhist propagandists were wise enough to suppress or keep out of sight all that might be offensive in their system, and to teach only such forms of doctrine as could be made attractive to the masses of the people. Kukai thus succeeded in converting the Mikado to his new interpretations of the Shinto beliefs, and the new system thus put forward received the name

NINJA STYLE

"Riyobu Shinto," which means "two parts/' or the "double way of the deitys," or the twofold divine teaching.

So complete and general did this Riyobu Shinto become in its spread throughout Japan that for a thousand years it dominated the civilization of the Empire. It had its priests, its gorgeous temples and ritual services, its philosophy, and its divers sects, and it is said that there are at least twelve distinct Buddhist sects in Japan to-day. The religion of the Buddha brought to Japan another and a wider humanizing influence a new gospel of tenderness together with a multitude of new beliefs that were able to accommodate themselves to the old, in spite of fundamental dissimilarity.

NINJA STYLE

In the highest meaning of the term, it was a civilizing power. Besides teaching new respect for life, the duty of kindness to animals as well as to all human beings, the consequences of all present acts upon the conditions of a future existence, the duty of resignation to pain as the inevitable result of for gotten error, it actually gave to Japan the arts and the industries of China. Architecture, painting, sculpture, engraving, printing, gardening in short, every art and industry that helped to make life beautiful developed first in Japan under Buddhist teaching."

Most of the Ninja practiced Buddhism and incorporated various ideas and views from Confucianism in to it. In addition, the Ninja prayed

NINJA STYLE

to various deities and Japan was covered with shrines and temples for this purpose. Some of them are dedicated to:

1. Fukurokujiu, deity of longevity, distinguished by a big head, long beard, a tortoise and a stork.

2. Hotei, deity of contentment, a dear old fat deity, whose delight is in playing with children.

NINJA STYLE

3. Daikoku, deity of riches, whose miner's hammer and bales of rice suggest the sources of Japan's wealth.

4. Yebis, deity of daily bread, especially fish; this being a favorite food in Japan.

5. Bishamon, deity of prosperity and military glory, usually dressed as an armed warrior.

6. Juro-jin, a very learned-looking figure, sometimes looked on as another form of Fukurokujiu.

7. Benten, deity of love, generally wearing a diadem and rich trailing robes.

NINJA STYLE

Many of the Ninja assumed the roles of religious pilgrims or religious beggars as a disguise. They may have only carried a simple 5 or 6-foot staff with them during their surveillance missions. The Ninja emphasized harmony with the 4 elements of earth, water, air, fire and brought these in to their beliefs. They had created complex charts that showed the Western astrological equivalents of Japanese ones. They believed in the curing properties of the earth and its ability to manufacture healing plants and herbs. The Shugendo religion pre-dated Buddhism and Confucianism in Japan and provided the basis for the beliefs of the Ninja, whether they subscribed to Buddhism, Confucianism, Hinduism, or Shinto-

NINJA STYLE

ism. The Ninja, coming from Ronin (Masterless Samurai) and Mountain Ascetic Priests (Yamabushi) took on the role of pilgrims as their open disguise in most instances. They were not pretending to play the role of religious pilgrims' despite being in disguise. The Ninjas believed in the role of religion in the success of their missions and they believed that they could receive divine protection through the use of invocations. The Ninja culture was not homogenous but the various clans did in fact share varying similarities in the way they practiced their beliefs.

NINJA STYLE

SIGNALLING

The Ninja maintained various forms of signaling so as to prevent enemy Ninja from infiltrating them. These signals allowed for the Ninja to know each other during an infiltration or escape. Many times during an escape, an enemy Ninja could infiltrate a Ninja group, giving away

NINJA STYLE

their positions and allies. The use of Ninjas would allow the enemy Ninja to not discover that he has been discovered. Signaling could be as silent and effective as hand signals to as complex and drawn out as conch shell codes. Conch shell codes could allow for a lower ranking to communicate with a higher ranking Ninja from a far distance, without creating a stir or commotion among castle defenders. Signals were well though, planned beforehand, and practiced in detail so there would be no mistakes in their use. The Ninja clan developed their own unique signaling systems and Ninjas from a particular clan had no connection in communication or organization with another Ninja clan. Each Ninja clan was unique in using

NINJA STYLE

their signaling systems to develop a specific advantage on a mission.

"You should write your name on your enemy's castle so everyone will know that it was you that achieved success."
- Bansenshukai

THE NINJA CULTURE

The Ninja were not one homogenous culture but they shared similarities in their beliefs. The Ninja came from an ascetic priest background or from masterless Samurai (Ronin). The Ninja took their belief in Shinto and Buddhism seriously and used Buddhist mantras to protect them on missions. The Ninja culture was a

NINJA STYLE

secretive one where the family living with the Ninja would not know that the person was a Ninja. Most of the time the Ninja were in gear that would make them seem like ordinary townspeople. Anyone caught walking around in such a uniform could be arrested, interrogated, and executed.

NINJA STYLE

The uniform was used for the theater and the sight of someone wearing one in the open would seem odd. Ninja practiced Shugendo and carried a deep tradition of protecting and connecting with the environment. The Ninja were very environmentally conscious and steered away from causing harm to the environment. They used the environment to heal themselves when hurt by picking plants that could provide them drinkable ointments. The Ninja also used the environment as their place of refuge during infiltration and during an escape. Their ability to disappear at will was because of their superior understanding of light and how it reflects. They were able to blend in with their environments because of

NINJA STYLE

their tactics as well as their flexible belief system that promoted harmony over other values. The Ninja were ruthless in their completion of missions but they believed that the art of Ninjitsu should be used for good. Whether this belief in fact influenced their behavior is another matter entirely as the Sengoku Jedai period contained times of both compassion and cruelty. An entire village would be slaughtered by a warring Daimyo looking to gain control so an assassination by a Ninja that could prevent thousands of deaths would be viewed as just and honorable. The belief in religion empowered Ninja to shed their fear on a particular mission and to use the powers through invocation to allow them protection. The use of

NINJA STYLE

Kuji Kiri was proof of their belief in supernatural forces to be able to protect them. This gave them a

reputation of being not only immensely fearsome in battle because of their skill sets but also because of their beliefs. Yamabushi or Warrior Monks sat in prayer and spent their time in practice, making

NINJA STYLE

them difficult to defeat. Their culture was one of honesty between each other and working towards achieving a pure mind. Their discipline gave them an advantage over the gluttonous Samurai class that sought out wealth, riches, and land. The Ninja class promoted frugality and simple living.

NINJA CRAZE

In the 1980's, Ninja movies were in vogue and production houses sought to give ninja themes to shows and movies. Toy makers chased after the creation of ninja related characters for their toy lines. The Ninja craze was becoming widespread and everyone sought to cash in. Ninja schools started propping up across the United States with so called

NINJA STYLE

Ninjitsu masters claiming direction connections with Japan. People began buying Ninja uniforms and Ninja weapons, pretending to be students of an art that had disappeared for hundreds of years. Movies like Enter the Ninja, Revenge of the Ninja, and American Ninja became huge hits. Movies like Bloodsport came out based on the life of a supposed Ninja master. The commercialization of anything Ninja related was mainstream and applauded. It is silly to think that in modern times the existence of Ninjas is a reality, but their techniques were in fact adapted by modern Special Forces units in the military. Ninjitsu never achieved the level of popularity that it did in the 1980's but it still remained an important art for

NINJA STYLE

study. Anyone claiming to be a Ninja
Master in this modern age is doing
so to demonstrate the ancient art. It
is more of a show at this point
because the art really stopped being
practiced nearly 400 years ago. The
techniques of the Ninjas remained
however and were adopted by
military units around the world. Ninja
were the most advanced warriors on
earth and that is due to their superior
training ethic. They spent thousands
of hours honing their skills so as to
be perfect on a mission. Their
livelihood and their life depended on
being able to complete a mission
with perfection. The Ninjas would
their time in silent practice. They
maintained their silence in practice
as well as in disguise. Their secrecy
is what made them so difficult to

NINJA STYLE

spot. But what also gave the Ninja a clear advantage is their use of the deadly skills of Ninjitsu as a last resort rather than as a first option.

DO NINJAS EXIST

The Ninjas have ceased to exist officially since the early 17th century but that has not stopped countless individuals from claiming that they own the secrets of Ninjitsu.

Countless movies were created in the 1980's themed around the Ninja narrative, books were written by supposed Ninja masters that were awarded their certification in Japan, and new cartoons like Teenage Mutant Ninja Turtles commercialized the word Ninja. Ninjas were featured in video games and nearly every major martial arts magazine started

NINJA STYLE

to feature articles related to the art of
Ninjitsu. Whether they refer to it as
Ninjutsu or Ninjitsu, they are talking
about an art that has ceased to
officially exist since 1603. Are these
authentic or in-authentic and why
does that matter? The authenticity of
a modern Ninja master is difficult to
be proven. Any ranking system or
belt system that is used in Ninjitsu is
obviously an innovation that was
borrowed from Judo and Karate.
Many so called modern Ninja
masters claim lineage to Ninja
Grandmasters, but any connection (if
indeed there is one) is at best
indirect and at worst a lie. Even
famous TV channels feature
documentaries showing supposed
"last remaining Ninja masters" but
any knowledge they have could only

NINJA STYLE

be theoretical. Did they use Ninja skills to assassinate people? Then they will be convicted of murder. In this modern age, assassination (whether silent or not) is called murder and it is a crime that is punishable by death (depending on the nation you are in). It is difficult to understand how any modern Ninja master could be a Ninja master if they have not used the skills of Ninjitsu (or Ninjutsu if you prefer) in actual real life situations. The closest thing to modern Ninjas could be military Special Forces. They are the closest thing to a modern Ninja because they both use stealth techniques to infiltrate an area and they use weapons to assassinate their enemies. The skills of the Ninja are as valid today as they were 400

NINJA STYLE

years ago, but its practice has evolved in to that of use for Army operations. There are no Ninjas walking around dressed in black throwing Ninja stars (shuriken) in this modern age. The skills of the Ninja were incorporated in to the activities of various Special Forces. Modern Special Forces utilize skills like stealth infiltration and escaping techniques that are based directly on the techniques of Ninjitsu. The SWAT teams of various Police Departments also incorporate the stealth techniques of Ninjitsu in training exercises used to better develop trainees for high pressure scenarios. Ninjitsu offered modern military and Police advanced techniques that would have taken

NINJA STYLE

them many years to develop on their own.

If you read the back of any martial arts magazine, you will see advertisements for videos by so called Ninja masters. If Ninjitsu hasn't existed in 400 years, how can it still be taught? The theory of the practice of Ninjitsu may be taught, but it is just in theory. If there are any legitimate practicing Ninja in this modern age, then it would mean that they are most like assassins. Ninjitsu, however you present it, is still a deadly art of assassination that wreaked havoc on Japan at its height. It is highly unlikely that there are paid assassins teaching their art through mail order. Ninjitsu, may be sold as a form of hand to hand

NINJA STYLE

combat, but it is really a form of armed combat. Tai-jutsu is the unarmed style of the Ninja. Are the books advertising Ninjitsu teaching Tai-Jitsu (Jutsu) or are they claiming to hold 400 year old secrets that they wish to transmit? Ninjas, as an organized unit, have ceased to exist for a long time and there inclusion in popular culture has kept their story alive.

There are various TV shows that feature interviews with living Ninja masters, showing their so-called skills in cutting a pineapple in half or using weapons to destroy various fruit. Cutting a pineapple in half proves nothing more than you are skilled in cutting pineapple in half. Ninja theory can be adapted and

NINJA STYLE

practiced in theory, but unless there are active individuals carrying out Ninja activities (infiltration, assassination, extraction), there words are most likely pure theory. Then again, if there are individuals carrying out Ninja activities (infiltration, assassination, extraction), then they would not be advertising to the general public to make a quick few dollars off them. Any modern association of Ninjas is purely indirect at best and make-believe at worst. Even if Hattori Hanzo's descendants practiced Ninja today, they would have no actual experience in carrying out Ninja activities and there knowledge would be entirely theoretical. Ninjas have ceased to exist in actual form for over 400 years.

NINJA STYLE

WAS NINJITSU CHINESE IN ORIGIN?

There is the mention of spies in Sun Tzu's Art of War but the use of individuals for assassination purposes was not mentioned by the famous Chinese General. Since Sun Tzu had been widely outspoken about the techniques of the Art of War, it would be natural and complimentary that the use of assassins would be pointed out and clearly evident. The claim that Ninjas have descended from China is not an accurate claim. It is speculative at best. Were the Ninjas taught by Chinese that entered Japan? They may have been taught by Chinese warriors that specialized in guerrilla warfare and this knowledge transferred over in to Japan, either

NINJA STYLE

through fighting or via independent teachers. Adaptation of techniques from China would not make Ninjitsu Chinese, but there is proof that shows Chinese did help the development of Ninjitsu by teaching guerrilla warfare techniques to Japanese warriors. But Ninjitsu is not just Guerrilla Warfare. Ninjitsu is a combination of Guerilla Warfare and Illusion, refined and polished in to a workable system that is both highly effective and highly deadly. The Japanese were surely influenced by Chinese warriors and learned from them, but this would not make Ninjitsu Chinese.

The art of Ninjitsu was invented in its modern form in Japan but it was its spreading to Western shores that gave it international notoriety.

NINJA STYLE

Guerrilla Warfare was the main doctrine that made up Ninja practices and it is indeed true that at least one Chinese master of warfare passed on knowledge to Japan but the practice of Ninjitsu is still uniquely tied to the Japanese culture. If the art of Ninjitsu was Chinese in origin, it would have been talked about by Chinese Generals like Sun Tzu in key military manuals like The Art of War. The fact that it was not mentioned is further proof that Ninjitsu was a Japanese invention that was unique to the times in which it was created. The Japanese are attributed with being the inventors of Ninjitsu because Ninjitsu was perfected in its current form in Japan. It is true that the Ninja have been attributed with mythical

NINJA STYLE

powers and that is due to the ignorance of the individuals that witnessed the exploits of Ninjas. Warriors attacking a Ninja compound would experience hundreds of Ninja disappearing in to virtually thin air. This was not because the Ninja possessed such powers, but rather that the Ninja were able to engineer and build clever mechanisms that allowed them to escape from their enemies under seemingly impossible circumstances. Ninjas are attributed with being able to disappear in a battlefield. Their opponents had little or no understanding of the advanced use of pyrotechnics for battle purposes. Ninjas advanced knowledge of the principles of illusion allowed them to blend in to an environment, thereby

NINJA STYLE

disappearing from the sight of their enemy. Chinese rarely ever used Ninjitsu tactics but it was the Japanese that adopted Guerrilla Warfare to create Ninjitsu. Without the knowledge of Guerrilla Warfare being passed from the Chinese to Japan, the art of Ninjitsu may have never been created. But it was the refining and polishing of the techniques by the Japanese that made the combination of illusion and guerrilla warfare, extremely difficult for an opponent to contend with or even to defend against. Ninjitsu reached its height of notoriety during the most brutal and bloody time in Japan, the Sengoku Jedai period. But once the period ended, the rule of law and political sophistication made the Ninja extinct.

NINJA STYLE

IS NINJITSU OBSOLETE

If it is not obsolete, it is at least not applicable. The use of the techniques of the Ninja were relevant to the time in which they were created. A time of turmoil and internal strife in Japan lead to the creation of Ninjitsu. The conditions existed for the creation of Ninja. The need for Ninja came out of a need to solve problems quickly and efficiently without the need to mobilize armies long distances. If a Daimyo wanted to punish a vassal, then it would invade that area with samurai, peasant soldiers armed with spears, and riflemen (if they could afford it). The entire venture was not only costly but more importantly time intensive. An immense amount of resources would have to be gathered

NINJA STYLE

and large numbers of troops mobilized to occupy an area. The more native the members of the Clan were defending the area, the more fiercely they would fight, as in the case of Warrior Monks. Warrior Monks fought harder than other Samurai, not because they wore Shrines around their neck, but because they were defending the local village and Monastery in which they lived and worked in. It was natural for Ninja, who came from ascetic warrior monks, to fight more fiercely. Their belief system empowered them but their sense of honor in defending the local village empowered them more. Their ways of connecting and harmonizing with nature are relevant to today's modern life, were polluting and living

NINJA STYLE

in pollution are the norm. The Ninja
believed that their connection with
the environment gave them strength,
allowing them to perform what others
would deem to be almost
supernatural. The Ninjas were able
to blend in to their environment with
ease because they understood the
principles of light and the creation of
the shadows. They blended in with
the shadows and used trees and
shrubs to easily hide from enemies.
Their ways have obviously been
adopted and influenced by designers
during the creation of camouflage.
Their advanced understanding of the
science of illusion allowed them to
avoid lighted areas and to blend in
with their environment with ease.
The Ninjas, whether they were
Buddhist or Shinto, lived a clean

NINJA STYLE

ascetic life. They avoided excess and extravagance, They practiced frugality and shared amongst each other. Many of the mountain ascetic monks became Ninjas and thus influenced the Ninja schools to invest in spiritual thought and practices. The Ninja were able to exist for long periods of time in complete quiet and isolation. This was due to their repeated sessions of meditation and sitting in silence. This, they believed, gave them the ability to read thoughts (telepathy) and to connect with their environment. This ability to remain quiet, allowed them to carry out long missions, sitting outside an enemy castle for long hours listening with focus. The Ninja did not suffer from attention deficit disorder rather they benefitted from Attention

NINJA STYLE

Surplus. They were able to maintain long periods of focus and their attention to detail gave them a definite advantage on the field of battle. Ninjas advanced use of multiple tools to achieve their objectives made them highly superior in combat operations. Tools for climbing, tools for moving across water, and tools for listening, made them able to capture information on their enemies without being detected. They were able to infiltrate an enemy castle with no trace of their entry, complete their mission, and escape without creating attention. This skill made Ninjas highly paid assassins for the various Daimyo looking to defeat their opponents using unconventional means of warfare. The Ninja clans

NINJA STYLE

would work for the highest bidder
because their skills could not be
duplicated by Samurai in armor
carrying multiple heavy weapons. A
Ninja could enter an enemy castle as
a worker and use a blowgun made
from bamboo to take out an enemy
Daimyo. A Samurai approaching that
same castle would be seen from far
away and would be stopped before
they even got close. Ninjas were, at
that time, the most highly advanced
soldier on earth, similar to military
Special Forces. Their use of multiple
tactics and special skills made them
unrivaled in a field of battle, though
they were a surgical strike rather
than a blunt instrument of warfare.
Ninjitsu depended on pinpoint
accuracy for reaching its goals while
Samurais hacked their enemies to

NINJA STYLE

death, making the battlefield look like a virtual bloodbath. Ninjas had the ability to end a battle decisively while preventing thousands of soldiers from killing each other on the battlefield. Ninja clans were hired by Daimyo to assassinate lower vassals and Daimyo that rivaled them. If a Ninja was caught, it meant a most embarrassing situation for the Daimyo that hired them. The Ninja were trained in many escape techniques, so as to stop them from being caught by the enemy. The Ninja could use water for hiding, the ground for disappearing under, trees to hide above, or objects to hide in (such as barrels). Having trained in escape techniques extensively made them naturals at escaping from situations that would make enemies

NINJA STYLE

think they possessed supernatural abilities. Ninjas will continue to excite readers and moviegoers alike, as their legend and the myths surrounding them increase. Their unique abilities to take out an enemy with minimal effort made them notorious and their highly advanced skill sets made them victorious. Their mannerisms and their ways are copied by modern martial artists that want to project a sense of connection with ancient Japan. Ninjas increased steadily in popularity from the early 1980's through books and movies focused around the Ninja fad. Artists include Ninjas in their artwork and game creators base games around them. Ninjas have been able to stay relevant because they are thought of

NINJA STYLE

as the forerunners and creators of a style that has been copied by modern Special Forces. Modern

CONCLUSION

Special Forces may deny their connection to ancient Ninjas, yet both Ninjas and Special Forces used advanced techniques of infiltration, advanced techniques of assassination, and advanced techniques of escape. The manuals of modern Special Forces read like ancient Ninja manuals, except that the spirituality has been removed from it. Ninjas have stayed relevant in modern times because their skill sets and techniques have stayed, to a certain extent, relevant in modern times. Their techniques have been adopted by military Special Forces

NINJA STYLE

around the world and this is why Ninjas are still the quintessential figure of war. Their stealth, their ability to disappear at will, to dispose of their enemies silently and quickly, to escape without a trace of their existence, made them highly valued and sought after. They worked for Daimyo that paid them the most but that did not make them un-loyal. The Ninja were highly loyal to the Daimyo that hired and they would swear and sign a written pact for their services, making their arrangement serious, sober, and business-like. They never betrayed their Daimyo and even risked their lives on multiple occasions to save Daimyo from certain destruction. During a time when Samurai were changing loyalty by the minute, Ninja were the most

NINJA STYLE

honest warriors in Japan. Being that the Ninja worked for money, they could have easily betrayed Daimyo, being In a position that was privy to information on other Daimyo. They did sell information to various Daimyo but still the Ninja did play favorites when given the opportunity. Certain Daimyo, which were more honorable, attracted Ninjas with similar goals and values. The Ninjas spent their time honing their skill sets so that they could attract the highest paying Daimyo. They operated as a business operation and were "for hire" fighters or in other words mercenaries. That is an over-simplification, because mercenaries don't tend to be honorable. During the Sengoku Jedai period, honor was preserved for storytelling and

NINJA STYLE

rarely did Samurai practice it. It was the ideal but not the norm. Samurai switches loyalties based on the amount of money they received and they switched loyalties when assigned official positions. Ninja did not seek official positions and did not attempt to work with the Shogun, as they preferred to operate independently of the central government and the various Daimyo (feudal landowners) ruling over Japan. The Daimyo were essentially warlords pushing on to each other for land and resources. The Shogun was powerless and was forced to work with the various Daimyo to maintain its own power. The Ninja were able to stay relatively neutral and work with various Daimyo for the

NINJA STYLE

purpose of achieving greater wealth and for unifying Japan.

The Ninjas are the stuff of legend and myth because their exploits were un-rivaled in world history. There ways allowed them to develop a style that made them valuable for the Sengoku Jedai period. The actual Ninja might be obsolete in a time of guns and laws, but their image of a mysterious warrior with nearly supernatural abilities continues to live in on the minds of readers. Ninjas continue to grow in popularity because they have become legendary. Ninjas will continue to excite us for as long as we are excited by the story of one person, against all odds, succeeding in their task.

NINJA STYLE

NOTES

NINJA STYLE

NOTES

NINJA STYLE

NOTES

NINJA STYLE

NOTES

NINJA STYLE

NOTES

NINJA STYLE

NOTES

NINJA STYLE

NOTES

NINJA STYLE

NOTES

NINJA STYLE

NOTES

NINJA STYLE

NOTES

NINJA STYLE

MIKAZUKI PUBLISHING HOUSE™
(U.S.P.T.O. Serial Number 85705702)

1) 25 Principles of Martial Arts
2) 25 Principles of Strategy
3) American Antifa
4) Arctic Black Gold
5) Art of War
6) Back to Gold
7) Basketball Team Play Design Book
8) Beginner's Magicians Manual
9) Boxing Coloring Book
10) California's Next Century 2.0
11) Camping Survival Handbook
12) Captain Bligh's Voyage
13) Coming to America Handbook
14) Customer Sales Organizer
15) DIY Comic Book
16) DIY Comic Book Part II
17) Economic Collapse Survival Manual
18) Find The Ideal Husband
19) Football Play Design Book
20) Freakshow Los Angeles
21) Game Creation Manual
22) George Washington's Farewell Address
23) Hagakure
24) History of Aliens
25) I Dream in Haiku
26) Internet Connected World

NINJA STYLE

27) Irish Republican Army Manual of Guerrilla Warfare
28) Japan History Coloring Book
29) John Locke's 2nd Treatise on Civil Government
30) Karate 360
31) Learning Magic
32) Living the Pirate Code
33) Magic as Science and Religion
34) Magicians Coloring Book
35) Make Racists Afraid Again
36) Master Password Organizer Handbook
37) Mikazuki Jujitsu Manual
38) Mikazuki Political Science Manual
39) MMA Coloring Book
40) MMA Dictionary
41) Mythology Coloring Book
42) Mythology Dictionary
43) Native Americana
44) Ninja Style
45) Ouija Board Enigma
46) Palloncino
47) Political Advertising Manual
48) Quotes Gone Wild
49) Rappers Rhyme Book
50) Self-Examination Diary
51) Shogun X the Last Immortal
52) Small Arms & Deep Pockets

NINJA STYLE

53) Stories of a Street Performer
54) Storyboard Book
55) Swords & Sails
56) Tao Te Ching
57) The Adventures of Sherlock Holmes
58) The Art of Western Boxing
59) The Book of Five Rings
60) The Bribe Vibe
61) The Card Party
62) The History of Acid Tripping
63) The Man That Made the English Language
64) The Whore Knows
65) Tokiwa
66) T-Shirt Design Book
67) U.S. Army Anti-Guerrilla Warfare Manual
68) United Nations Charter
69) U.S. Military Boxing Manual
70) Van Carlton Detective Agency; The Burgundy Diamond
71) William Shakespeare's Sonnets
72) Words of King Darius
73) World War Water

If the Mikazuki Publishing House™ book is not available, place a request with any bookstore to order it for you.
Instagram.com/MikazukiPublishingHouse

NINJA STYLE

KAMBIZ MOSTOFIZADEH TITLES

1. 25 Principles of Martial Arts
2. 25 Principles of Strategy
3. American Antifa
4. Arctic Black Gold
5. Back To Gold
6. Camping Survival Handbook
7. Economic Collapse Survival Manual
8. Find the Ideal Husband
9. Game Creation Manual
10. History of Aliens
11. Internet Connected World
12. Karate 360
13. Learning Magic
14. Magic as Science & Religion
15. Make Racists Afraid Again
16. Mikazuki Jujitsu Manual
17. Mikazuki Political Science Manual
18. MMA Dictionary
19. Mythology Dictionary
20. Native Americana
21. Ninja Style
22. Ouija Board Enigma
23. Political Advertising Manual
24. Small Arms & Deep Pockets
25. The Bribe Vibe
26. The Whore Knows
27. Van Carlton Detective Agency: Burgundy Diamond
28. World War Water

Facebook.com/KambizMostofizadeh